AF428253

Black Holes, White Holes, and Superluminal Starships

Stephen Blaha Ph. D.
Blaha Research

Second Kind White Holes
Generation of Holes
Variable Speed of Light
White Hole Starship Engines
Gravity Boosts for Superluminal Starships
White Holes and Universe Desert Regions
White Hole Colliding Beam Accelerators
Superluminal Extension of Newton's Second Law

Pingree-Hill Publishing
MMXXIV

Rev. 00/00/01   January 14, 2025

To Margaret

Some Other Books by Stephen Blaha

SuperCivilizations: Civilizations as Superorganisms (McMann-Fisher Publishing, Auburn, NH, 2010)

All the Universe! Faster Than Light Tachyon Quark Starships & Particle Accelerators with the LHC as a Prototype Starship Drive Scientific Edition (Pingree-Hill Publishing, Auburn, NH, 2011).

Unification of God Theory and Unified SuperStandard Model THIRD EDITION (Pingree Hill Publishing, Auburn, NH, 2018).

The Exact QED Calculation of the Fine Structure Constant Implies ALL 4D Universes have the Same Physics/Life Prospects (Pingree Hill Publishing, Auburn, NH, 2019).

Passing Through Nature to Eternity ProtoCosmos, HyperCosmos, Unified SuperStandard Theory (Pingree Hill Publishing, Auburn, NH, 2022).

HyperCosmos Fractionation and Fundamental Reference Frame Based Unification: Particle Inner Space Basis of Parton and Dual Resonance Models (Pingree Hill Publishing, Auburn, NH, 2022).

God and and Cosmos Theory (Pingree Hill Publishing, Auburn, NH, 2023).

Newton's Apple is Now The Fermion (Pingree Hill Publishing, Auburn, NH, 2023).

Cosmos Theory: The Sub-Particle Gambol Model (Pingree Hill Publishing, Auburn, NH, 2023).

Cosmos-Universe-Particle-Gambol Theory (Pingree Hill Publishing, Auburn, NH, 2024).

Fractal Cosmos Curve: Tensor-based Cosmos Theory (Pingree Hill Publishing, Auburn, NH, 2024).

The Eternal Form of Cosmos Theory Third Edition (Pingree Hill Publishing, Auburn, NH, 2024).

Geometric Cosmos Geometric Universe (Pingree Hill Publishing, Auburn, NH, 2024).

Particles and Universes of Cosmos Theory (Pingree Hill Publishing, Auburn, NH, 2024).

Unification of the Subluminal and the Superluminal in Cosmos Theory (Pingree Hill Publishing, Auburn, NH, 2024).

The Dawn of Dynamic Cosmos Dimension Arrays (Pingree Hill Publishing, Auburn, NH, 2024).

Structure and Dynamics of Cosmos Theory and the Unified SuperStandard Theory (Pingree Hill Publishing, Auburn, NH, 2024).

Available on Amazon.com, bn.com Amazon.co.uk and other international web sites as well as at better bookstores.

CONTENTS

FIGURES and TABLES

Introduction

In view of the maladies afflicting this planet isn't it time to look to the other planets and stars for the future. Here's a new view of the ongoing process in our solar system and the beginning of a look to a future in the stars. This book begins with a description of the needs and required methods for a satisfactory, economical program for travel to the planets of the solar system. The defects of a reliance on chemical rockets are presented. The necessity of a nuclear rocket based program is presented.

It then develops a proposal for the use of mini White Hole based starship engines and for gravity assists for starships. New solutions are presented for White Holes, and White Holes of the Second Kind, which are similar to the Schwarzschild solution for Black Holes. White Holes do not have an event horizon. White Holes of the Second Kind do have an event horizon.

The use of mini White Holes for colliding beam particle accelerators is discussed.

How do we create White Holes? The book suggests they may be generated by creating macroscopic quantities of the author's multiquark quarkium particles using inertial compression of tritium. Then the quarkium fluid may take the form of "Holes" that may be assembled into a White Hole engine or used in an ultra-high energy colliding beam accelerator.

The mechanics of superluminal motion are derived. A superluminal extension of Newton's Second Law of Mechanics is presented based on this study of superluminal motion.

This book is an endeavor to begin a serious study of superluminal starship travel.

1. A Rational, Cost-Effective Approach to Space Travel[1]

NASA, and the space programs of other countries, has done an admirable job in Near Space projects such as communications satellites, the space station, and space-based science projects. They have also sent exploration ships to the moon, Venus, Mars, and the outer planets and moons which have generated vast amounts of data that will help us understand planetary dynamics and the specific features of planets and moons. The study of Jovian and Saturnian moons with their possible subsurface oceans and concomitant possibility of alien life has aroused great public and scientific interest.

All of these efforts have been based on the use of chemical rockets. Chemical rockets have the advantage of a well-developed technology that has incrementally advanced since the 1940's. Unfortunately they also have significant disadvantages for long trips to other Solar System planets. The disadvantages are costliness and lengthy travel times. Lengthy travel times create major human factors issues for manned travel to even the closest planets. Other issues include atmospheric pollution by rocket exhaust, outer space radiation, psychological effects of long isolation, and the provision of food, water and oxygen for the astronauts.

Recently Russia, and the United States, has again[2] begun to consider the use of nuclear rockets for planetary voyages. Nuclear rockets can significantly reduce travel times to the planets thus reducing the negative effects listed in the previous paragraph. They also are necessary for manned voyage to the outer planets.

Recently, a private rocket launch industry in the United States has appeared. The companies in this fledgling industry are committed to the use of various forms of chemical rockets to launch payloads into Near Space, the moon and possibly beyond. Sadly they do not use nuclear rocket technology and so only offer a modest possible improvement in space travel

It is apparent to this author, and others, that these new efforts are attempts to develop a comprehensive, cost-effective space program for humanity. However, it is the opinion of this author that there are significant possibilities that have been overlooked in the present and emerging space program. The world-wide funding crunch has made it imperative to examine all reasonable, possible ways to develop a world space program in the most cost-effective manner while not sacrificing opportunities for the exploration of the Solar System and beyond. This book proposes a program that will achieve that goal if certain technological challenges are successfully overcome. It also suggests a

---

[1] Much of this chapter first appeared in Blaha (2013a).

[2] A nuclear rocket program was started and restarted and then abandoned by NASA's predecessor, the National Advisory Committee for Aeronautics (NACA). For example, see "Steady Nuclear Combustion in Rockets" by E. Sänger, Astronautica Acta, I, Fasc. 2 (1955).

provocative, *eventually possible* future for faster than light interstellar travel: Superluminal starships.

Since there are several aspects to space travel from the earth we have chosen to focus on the three major aspects of space travel.[3] We have also chosen not to discuss existing space vehicles, but rather to consider new, potential types of space transport bearing in mind the current, and probable near term economic problems which make it imperative to maximize the return on investment in the development and use of space transport vehicles.

1.1 General Requirements for Cost-Effective Spaceship Propulsion

There are self-evident requirements for spaceship propulsion methods at the present time and for the foreseeable future:

1. For a given type of route the propulsion method must be timely and not be slow in terms of human life spans.

2. The propulsion method chosen for a particular class of routes should be the most cost-effective method.

3. The propulsion methods currently possible for any set of routes should be evaluated to find the best propulsion method.

4. As our physical and engineering knowledge evolves over the years, reassessments of propulsion methods should be made for each set of routes. The continuing development of new materials and new energy sources makes periodic[4] reassessment of propulsion methods necessary.

Space travel routes can be divided into three types. Each type has one or more preferred methods of travel. We will consider these types of travel routes and their propulsion mechanisms below. Subsequent chapters will make more detailed proposals.

1.2 Three Phases of Space Travel

1.2.1 Near Space Propulsion

The first type of route is travel from the surface of the earth to near regions of space, which we can take to be 150 kilometers. Most space activity in the past forty years has concentrated on this phase with communication satellites being a primary purpose. Chemically powered rockets have been the only propulsion method used until the present.

While chemical rockets and shuttles have been very successful in accomplishing their missions, it appears that other propulsion methods may be more cost-effective for Near Space missions. Building and fueling large rockets is very costly.

---

[3] Travel and transport near earth, in the Solar System, and to the stars.
[4] Reassessments should perhaps be made every twenty years.

We suggest a suitably designed *multi-stage space gun* may provide a more cost-effective delivery vehicle for the delivery of most non-fragile cargo to Near Space. We have described this new type of space gun previously. Its ancestors are the German big guns of World War I. These guns of World War I (the approximately 100 foot long Big Bertha and a larger gun) that bombarded Paris from a distance of 80+ miles sent their 100+ lb. shells as high as 80 – 90 miles above the earth to NEAR Space. Their shells had a speed of 1 mile per second as they emerged from the gun's barrel.

The "single-stage" German big guns had a number of deficiencies as a method to loft cargo into space. Gun barrels eroded fairly quickly. More importantly their cargo capability did not scale upwards satisfactorily from one kilogram shells to very heavy shells. A one kilogram payload required at least 1 kilogram charge to propel into Near Space. However a much larger payload, for example a 100 kg payload, would require a much larger charge because the charge would not only propel the payload but it would also propel the "unburned" charge above it in the gun thus increasing the required propellant charge significantly.

A single or multi-stage rocket, with or without boosters, uses over 99% of its weight as fuel to send a payload into space. The fuel is, for the most part, used to propel itself (the fuel) off the ground at an ever-increasing speed into space. In contrast, the propellant for a gun propels the projectile and a fraction of the propellant between the point of burn of the powder charge and the projectile. Thus a sufficiently large single-stage gun could efficiently put a payload up eighty miles into Near Space because the propellant propels the payload and the charge, and not fuel or a rocket casing.

However, single-stage gun barrel erosion and the propellant required for large cargoes, among other factors, led NACA[5] not to pursue this technology in the 1960's.

Previously we described the design of a multi-stage space gun that would avoid the escalation of propellant charge needed for large payloads and reduce the barrel erosion problem. We view multi-stage space guns as a cost-effective solution to send bulk cargo to space while continuing to use chemical rockets for people transport and the transport of delicate equipment. The multi-stage space guns could enable the transport of large quantities of materials to space to build a large space station and to build (nuclear) rockets in orbit to avoid potential nuclear catastrophes due to the launch of nuclear rockets from earth. They also can be important in building the infrastructure of a spaceport on the moon and ultimately Mars. The most expensive part of these projects will be the initial stage of transport of massive quantities of materials to Near Space in a cost-effective manner.

1.2.2 Solar System Travel – Nuclear Rockets

A nuclear rocket program was started and then abandoned by one of NASA's predecessors, the National Advisory Committee for Aeronautics (NACA). It was felt that nuclear rockets were dangerous should they explode on takeoff from earth or from

---

[5] The NASA predecessor.

near earth orbit. In view of the Chernobyl nuclear reactor disaster which impacted on much of Europe this decision was prudent.

However if we can develop mass transport of nuclear rocket construction materials to far earth orbit, or better yet to orbit around the moon, by using methods such as multi-level space guns, then the dangers of nuclear rocketry could be reduced to a minimal level. After construction, nuclear rocket(s) could be used to explore the Solar System much more quickly and economically than chemical rockets. For example a chemical rocket trip to Mars is estimated to take approximately two years while a nuclear rocket trip would take approximately six months (and perhaps much less if the nuclear rocket was especially powerful.)

The cost of developing a nuclear rocket would be significant. But its use in interplanetary travel would be far more cost-effective then a chemical rocket. An additional economic benefit would be the reusability of the nuclear rocket for repeated voyages. Rather like nuclear submarines they offer great cruising ranges and long range trips.

Finally a well-designed nuclear ion rocket could use hydrogen, methane and other gases mined on the asteroids and moons on outer solar system planets as a source of ion particles. Thus a nuclear rocket need not carry large quantities of the propellant as opposed to chemical rockets which generally need to carry large amounts of fuel. Even if chemical fuels are found on distant asteroids they would, most likely, need to bring a large supply of oxygen.

We can only conclude that a top priority nuclear rocket R&D program is needed. Given tight budgets, expenditures on chemical rocket ventures should be diverted to the nuclear rocket program. Quite simply, we want the horse before the buggy. There is no good reason to expend large sums immediately for most of the science driven rocket trips currently planned. The Solar System will not change much in the next ten years. We should divert funds to a workhorse nuclear R&D program and wait until a nuclear rocket(s) can carry out the delayed projects expeditiously and economically. The Russian space program has recognized the importance of nuclear rockets and is actively engaged in their design and development. The US Space Program should pursue a similar course.

1.2.3 Extra-Solar Travel

There are two major propulsion issues for travel to other stars and galaxies. First a sensible propulsion method must be found that enables faster-than-light travel to stars and galaxies. When a starship reaches a distant star then a secondary propulsion system must be present for travel within the star's Solar System. The secondary propulsion system would be a combination of nuclear and chemical propulsion systems. The nuclear propulsion unit would handle large-scale movement within the remote solar system. The chemical propulsion unit would handle motion in the vicinity of a specific planet or moon. The starship should also have two nuclear shuttles for round trip visits to the surface of earth-like planets and the surface of moons.

1.2.3.1 Sub-Light Starships

Many proposals have been made for sub-light travel to the stars. One type of proposal suggests a nuclear powered ion drive starship. More exotic proposals exist such as solar sails that use starlight to accelerate a starship. All of these sub-light starship proposals require extremely lengthy travel times extending up to many generations. As a result they have a very limited range of nearby stars and secondarily they are not suitable for the important goals of mining, trade and colonization should we find planets/moons with important minerals, or aliens with whom to trade, or planets similar to earth suitable for human colonization.

Thus these types of starships will only be useful for scientific exploration. This is a worthy goal. But the possibilities of the previous paragraph would appear to be far more significant for the long-term benefit of Mankind. We therefore believe sub-light starships are not desirable.

1.2.3.2 Faster-Than-Light Gravity-Propelled Starships

There are a number of proposals for starship propulsion based on gravitation effects. A prominent propulsion proposal is called *Alcubierre Drive*. It was widely discussed at a recent conference on starships. However the development of such a drive is far beyond the capabilities of Humanity for the foreseeable future – thousands of years. A major reason for its practical impossibility is that it requires the manipulation of masses of the size of Jupiter – utterly impossible. Can one imagine a "trick" that would make Alcubierre Drive, or any other gravitational drive, feasible. General Relativity gives a resounding no. General Relativity has been a subject of study for almost 100 years. *It is clear that the weakness of the gravity force, and thus the need for extremely large masses to have large gravitational effects, makes gravity driven starships impossible.*

A new possibility presents itself in this book. If we could develop starship engines based on mini Second Kind White Holes then the thrust could be superluminal and a starship might achieve superluminal speeds – a Gateway to the stars. See chapter 7.

1.2.3.3 Faster-Than-Light Starships

There are many reasons why starships must go faster-than-light. Some of them have been discussed earlier. The primary reason can be seen in the history of world transportation. The trend of trade and migration between countries grew enormously as transportation became faster and vehicles became larger and safer. The transition from sailing ships to steamships to aircraft has had a large effect on shaping the present world.

As we reach out into the Solar System, and then to the stars, the speed and capacity of transportation must grow significantly.

1.2.3.4 Nuclear Rockets for Travel in Other Star Systems

Because of the nature of faster-than-light travel for distances of hundreds or thousands (or more) light years, nuclear rockets and chemical rockets for use at a distant destination must have an extremely long "shelf" lifetime – perhaps of the order of millions of years for especially long distance travel to other galaxies. They would rest unused during a long starship time period and, upon arrival at a remote destination, they would be put into use. Long life nuclear reactors for nuclear rockets are yet to be developed. Long life chemical rockets have the problem of field deterioration.

1.3 Space Program Prioritization

The current approach of the space-faring nations is to use chemically powered rockets. Make bigger rockets to explore further in the solar system – particularly manned expeditions, and to build colonies in space, and on the moon and Mars.

The United States and Russia have also begun exploring an alternate path to space using nuclear powered rockets. Nuclear rockets have some decided advantages for longer trips in the solar system. For example a trip to Mars via chemical rockets takes about two years. A nuclear rocket could make the same voyage in about six months.

The United States has indicated preliminary interest in restarting a nuclear rocket program of the 1950's and 1960's.[6] Whether these ventures are brought to fruition, or not, they represent a fresh approach that would of great benefit for space programs.

The only satisfactory way to explore the solar system is using nuclear rockets constructed in earth orbit initially and eventually in moon orbit. The construction process would be best done with space guns that can send large payloads into orbit economically (and in a less environmentally damaging manner – chemical rockets are polluters.)

1.4 A Starship Possibility

This book presents a new mechanism for starship propulsion based on mini Second Kind White Hole engines. It also suggests a Second Kind White Hole gravity boost mechanism while starships are in transit amongst the stars.

---

[6] A nuclear rocket program was started and then abandoned by NASA's predecessor, the National Advisory Committee for Aeronautics (NACA). For example, see "Steady Nuclear Combustion in Rockets" by E. Sänger, Astronautica Acta, I, Fasc. 2 (1955)

2. Schwarzschild Solution and the Speed of Light

Many mechanisms for travel to the stars have been proposed. We will later suggest a new mechanism for interstellar travel that offers the possibility for faster than light propulsion. It is based on a new solution of the Einstein metric equations. It is related to the Schwarzschild metric that is used for Black Hole studies. It supports a speed of light in its vicinity that may be substantially in excess of the vacuum speed of light c.

A starship can massively accelerate to velocities far in excess of c and travel expeditiously to the stars.

We begin by examining the speed of light, which may vary near Holes, and then progress to consider the speed of light in an alternate solution of the Einstein equations.

2.1 Known Schwarzschild Metric and Black Holes

The Schwarzschild metric proper time interval is

$$ds^2 = c^2 d\tau^2 = c^2 A_t dt^2 - A_r dr^2 - r^2 d\theta^2 - r^2 \sin^2\theta\, d\varphi^2 \qquad (2.1)$$

where M is the mass, G is the gravitational coupling constant, and

$$A_t = (1 - 2MG/rc^2)$$
$$A_r = 1/(1 - 2MG/rc^2) \qquad (2.2)$$

In the vacuum case where M = 0 the metric is

$$ds^2 = c^2 d\tau^2 = c^2 dt^2 - dr^2 - r^2 d\theta^2 - r^2 \sin^2\theta\, d\varphi^2 \qquad (2.3)$$

Its time and radial distance intervals coefficients give the speed of light squared c^2.

However in the case of the Schwarzschild metric the effective speed of light near a Black Hole, based on the coefficients of the time and radial intervals is

$$v_c^2 = c^2 A_t/A_r = c^2(1 - 2MG/(rc^2))^2 \qquad (2.4)$$

2.1.1 Relation of Speed of Light and Black Hole Particle Entry

The effective speed of light near a black hole is:

$$v_c = c(A_t/A_r)^{1/2} = c(1 - 2MG/(rc^2)) \qquad (2.5)$$

As $d = 2MG/(rc^2) \rightarrow 1$ (from the region exterior to the Black Hole event horizon) we see $v_c \rightarrow 0$. At the Black Hole event horizon, $v_c = 0$. In the interior of the

Black Hole v_c becomes imaginary and progresses from 0 to $i\infty$. The Physical implication of $v_c = 0$ is that it represents the transition of particles from bradyons to tachyons within the event horizon.

Within the event horizon, the time, t, and radial distance, r, are interchanged. Correspondingly particle energy and radial momentum are interchanged causing the particle mass^2 = energy^2 – momentum^2 to become zero at the event horizon and become negative (tachyonic) within the event horizon.

External to the event horizon v_c ranges from 0 to c as r becomes large.

2.2 Application to Starship Propulsion

The above Schwarzschild solution does not provide a gravity boost to starships. In the vicinity of a black hole the effective speed of light is lower than its speed in the vacuum. As a result the passage of an accelerating starship through the vicinity of a black hole will *diminish* the acceleration of the starship due to the effect of the $\gamma = (1 - v^2/v_c^2)^{-\frac{1}{2}}$ factor.

On the other hand, a larger v_c value would *enhance* the acceleration of the starship to higher velocities by diminishing the $\gamma = (1 - v^2/v_c^2)^{-\frac{1}{2}}$ factor, which is effectively a *hindering* effect dynamically.

In chapter 4 we will develop a "hole" model that enables the speed of light v_c to become large in the neighborhood of the hole. As a result a starship will have an enhanced "jump" to higher velocities in this region – a gravity boost.

3. A White Hole Solution Without an Event Horizon

3.1 White Hole Solution

The Schwarzschild black hole solution with an event horizon was a source of concern to physicists in the 1930s and 1940s. The discovery of black holes removed that concern. However the possibility of similar solutions without an event horizon is of interest since it has potential experimental astrophysical possibilities.

In this chapter we present a new general relativistic solution that has features similar to the Schwarzschild solution but *without* an event horizon. Consequently one may "see" the innards of this type of gravity hole. We call it a *White Hole* on that basis.

We now consider a General Relativistic solution of the field equations near a large mass which will be seen to *not* have an event horizon.

Following a thought process similar to that which leads to the Schwarzschild metric we begin with a new time interval expression:

$$ds^2 = c^2 d\tau^2 = c^2 A_t dt^2 - A_r dr^2 - r^2 d\theta^2 - r^2 \sin^2\theta \, d\varphi^2 \qquad (3.1)$$

where A_t and A_r are functions of r.

The Ricci tensor with $R\mu\nu = 0$ implies

$$A_r A_t = const \qquad (3.2)$$

We require A_t and A_r approach Minkowski metric values at large distance:

$$A_t \rightarrow 1 \quad A_r \rightarrow 1 \qquad \text{for large r} \qquad (3.3)$$

which implies

$$A_r = 1/A_t \qquad (3.4)$$

We require the Ricci tensor component $R_{\theta\theta} = 0$, which implies the condition:

$$r \, dA_t/dt + A_t = 1 \qquad (3.5)$$

or

$$dA_t/dt = (1 - A_t)/r \qquad (3.6)$$

If we set (Note: this is contrary to the Schwartzschild solution)

$$A_t = 1/(\, 1 + 2MG/c^2 r)$$
$$A_r = (\, 1 + 2MG/c^2 r) \qquad (3.7)$$

then

$$dA_t/dt = (2MG/c^2r^2)/(1 + 2MG/c^2r)^2 \qquad (3.8)$$

For large r the metric component g_{tt} becomes

$$g_{tt} = -c^2A_t \qquad (3.9)$$

We require g_{tt} must approach Newtonian gravity:

$$g_{tt} \approx -c^2 - 2\varphi \qquad (3.10)$$

where $\varphi = -MG/r$. Then

$$A_t \rightarrow c^2 + 2\varphi = c^2 - 2MG/c^2r \qquad (3.11)$$

is required. Eq. 3.8 for large r :implies:

$$dA_t/dt \rightarrow 2MG/c^2r^2 \qquad (3.12)$$

The right side of eq. 3.6 is

$$(1 - A_t)/r = (1 - 1/(1 + 2MG/c^2r))/r \qquad (3.13)$$

Expanding eq. 3.13

$$(1 - A_t)/r \approx 2MG/c^2r^2 \qquad (3.14)$$

Comparing eqs. 3.12 and 3.14 we find eq. 3.10 is satisfied, thus verifying eqs. 3.7. Thus we have a new solution for a large mass M.

3.2 White Hole Singularity

   The positivity of the terms in eqs. 3.7 eliminate the possibility of an event horizon. Thus we have a White Hole allowing its interior to be directly viewed. Whether there are White Holes in the universe is at present an open question. There are some experimental suggestions of White Holes.

3.3 Relation of Speed of Light and the White Singularity

   The speed of light in the vicinity of the White Hole is

$$v_c^2 = c^2A_t/A_r = c^2(1 + 2MG/rc^2)^{-2} \qquad (3.15)$$

As $r \rightarrow 0$ we find $v_c \rightarrow 0$. For large r we find $v_c \rightarrow c$. As a result this new solution does not have significant consequences for starship motion. In the next chapter we consider a further variation that might affect/improve starship dynamics.

4. White Holes of the Second Kind with an Event Horizon

There is a general belief that gravity is always attractive. However possible modifications of gravity at large distances such as MOND and this author's studies of higher derivative gravitation[7] suggest that the possibility gravity may be in part repulsive – particularly at large distances.

There are also regions in the universe that are seemingly devoid of matter – deserts. The manner in which these regions were created and how they are maintained remains an open question. Their existence raises the question that they may be the result of a repulsive form of gravity residing within them that maintains a desert of galaxies. The mechanism may be a population of a type of White Holes with event horizons, White Holes of the Second Kind, that embody a repulsive form of gravity.

Chapter 6 points out that forces, and their potentials, change from repulsive to attractive when going from subluminal to superluminal motion.

This chapter derives this form of White Holes of the Second Kind. After formation they exhibit a repulsive gravitation that repels matter from falling into them. They have an event horizon the blocks viewing their contents. They are a variation on Black Holes.

In the neighborhood of such a White Hole the speed of light will be seen to exceed the conventional speed of light. As a result starships traversing this region will have an enhanced acceleration (gravity boost) that will expedite their travel to distant regions. The velocity of light in the vicinity of a Second Kind White Hole is

$$v_c^2 = c^2 A_t/A_r = c^2(1 - 2MG/(rc^2))^{-2} \qquad (4.1)$$

where M is the mass of the White Hole, G is the gravitational constant and r is the radial distance from the center of the Second Kind White Hole. Due to the appearance of the minus sign in eq. 4.1 we see as $r \rightarrow 0$ the speed of light increases without limit. Consequently the γ factor that appears in dynamical equations becomes effectively one and thus acceleration is not so limited. We see starships having a form of gravitational slingshot traversing the environment of a Second Kind White Hole.

4.1 Derivation of White Holes of the Second Kind
This derivation follows the form of the previous derivations with significant changes. The most important change is using a repulsive gravity potential of the form:

$$\varphi = +MG/r \qquad (4.2)$$

---

[7] See for example Blaha (2018e).

in the derivation. *This change is consistent with the change from repulsive to attractive in the effective force and potential for superluminal dynamics. See chapter 6.*
    We begin with the time interval expression:

$$ds^2 = c^2 d\tau^2 = c^2 A_t dt^2 - A_r dr^2 - r^2 d\theta^2 - r^2 \sin^2\theta \, d\varphi^2 \qquad (4.3)$$

where A_t and A_r are functions of r.
    The Ricci tensor with $R\mu\nu = 0$ implies

$$A_r A_t = \text{const} \qquad (4.4)$$

We require A_t and A_r approach Minkowski metric values at large distance:

$$A_t \to 1 \quad A_r \to 1 \qquad \text{for large r} \qquad (4.5)$$

which implies

$$A_r = 1/A_t \qquad (4.6)$$

We require the Ricci tensor component $R_{\theta\theta} = 0$, which implies the condition:

$$r \, dA_t/dt + A_t = 1 \qquad (4.7)$$

or

$$dA_t/dt = (1 - A_t)/r \qquad (4.8)$$

We now set

$$A_t = 1/(1 - 2MG/(c^2 r)) \qquad (4.9)$$
$$A_r = (1 - 2MG/(c^2 r))$$

The change in sign relative to the White Hole case of chapter 3 will be seen to be a consequence of the change from attractive gravity to repulsive gravity. Then

$$dA_t/dr = 2MG/(c^2 r^2)/(\, 1 - 2MG/(c^2 r))^2 \qquad (4.10)$$

For large r, the metric component g_{tt} becomes

$$g_{tt} = -c^2 A_t \qquad (4.11)$$

We require g_{tt} must approach *repulsive* Newtonian gravity (eq. 4.2):

$$g_{tt} \approx -c^2 - 2\varphi \qquad (4.12)$$

where $\varphi = +MG/r$. Then

$$A_t \to c^2 + 2\varphi = c^2 + 2MG/(c^2 r) \qquad (4.13)$$

is required. Eq. 4.10 for large r implies:

$$dA_t/dr \rightarrow 2MG/(c^2r^2) \qquad (4.14)$$

The right side of eq. 4.8 is

$$(1 - A_t)/r = (1 - 1/(1 - 2MG/(c^2r)))/r \qquad (4.15)$$

For large r, eq. 4.15 is approximately

$$(1 - A_t)/r \approx 2MG/(c^2r^2) \qquad (4.16)$$

Comparing eqs.4.14 and 4.16 we find eqs. 4.9 is satisfied, thus consistently verifying our choice of the form of eqs. 4.9. Consequently we have a new solution.

4.2 Velocity of Light Near White Holes of the Second Kind

The event horizon of White Holes of the Second Kind is at the radial distance:

$$r = 2MG/c^2 \qquad (4.17)$$

The speed of light in the vicinity of White Holes of the Second Kind is

$$v_c^2 = c^2 A_t/A_r = c^2(1 - 2MG/(rc^2))^{-2} \qquad (4.1)$$

As $r \rightarrow 2MG/c^2$, the event horizon, $v_c \rightarrow \infty$. For large r we find $v_c \rightarrow c$. As a result this new solution might have significant consequences for starship motion in the vicinity of a Second Kind White Hole.

In the following chapters we consider the impact of a much higher $v_c > c$ on starship motion. We find the acceleration to high velocities may be much enhanced – opening the possibility of a gravitational starship slingshot.

It also opens the possibility of a White Hole of the Second Kind – based starship engine that uses a "small" Second Kind White Hole(s) to produce thrust at superluminal velocities for a starship. We discuss this possibility later.

4.3 Starship Motivation for White Hole of the Second Kind

An increase in the effective speed of light causes an increased acceleration per unit fuel (force) as we will see later. As a result we see the following possible gravity assist from a Second Kind White Hole:

1. Go up to near starship velocity $v = c$
2. Enter vicinity of a Second Kind White Hole
3. Accelerate to higher velocity taking advantage of high v_c.
4. Exit White Hole vicinity - moving much faster than c.
5. Accelerate to even higher speed since we have gone beyond the $v = c$ singularity – now we are superluminal with no $v = c$ barrier (although there is still a $v = v_c$ singularity.)

The use of a Second Kind White Hole would facilitate superluminal starship travel. The problem is to be near one.

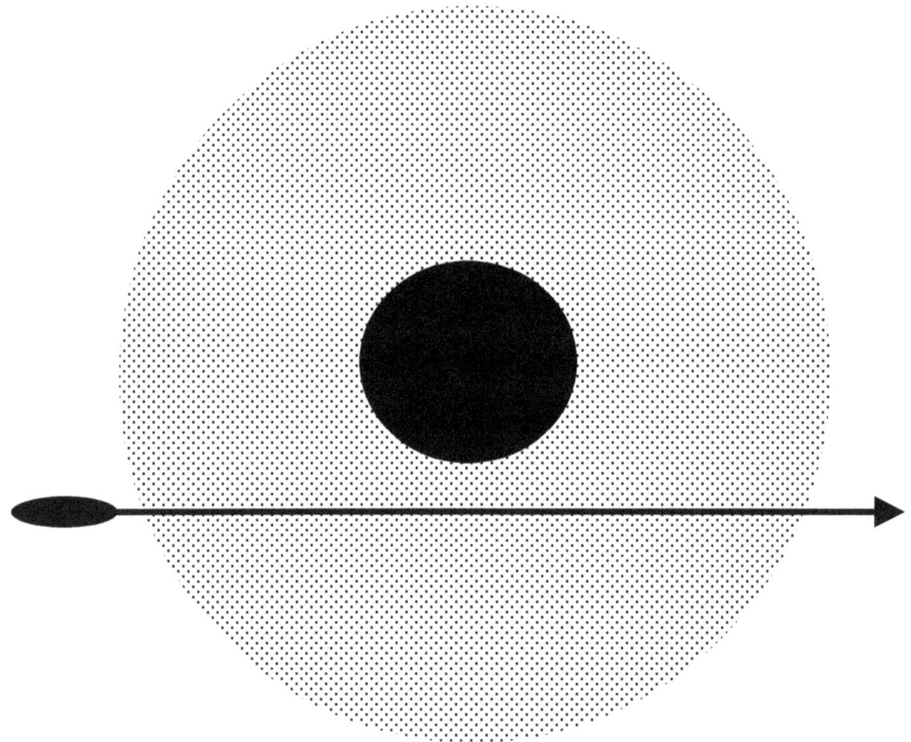

Figure 4.1. A starship passing near a Second Kind White Hole to get a gravity boost of acceleration.

5. Superluminal Dynamics

This chapter develops the dynamics of the motion of a particle or starship as a function of the speed of light which we will take to vary in different space-time regions. The previous chapter considered the effect of a Second Kind White Hole on the speed of light in its vicinity.

We consider the cases of relativistic dynamics for velocities less than the speed of light: $v < v_{s.}$, and for velocities greater than the speed of light: $v > v_c$. *We will see, for all else being equal, particles accelerate faster for large v_c whether they are moving faster than the current speed of light v_c or less than the current speed of light v_c. Thus starship acceleration could be significantly enhanced near a Second Kind White Hole.*

5.1 Dynamics for Velocities less than the Speed of Light

The standard non-relativistic expression for force and acceleration is:

$$F = md^2x/dt^2 \tag{5.1}$$

where m is the mass. In subluminal relativistic motion eq. 5.1 is generalized to

$$F^\mu = dp^\mu/d\tau \tag{5.2}$$

for momentum p^μ where

$$dt = \gamma d\tau \tag{5.3}$$
$$p^\mu = mdx^\mu/d\tau$$
$$v = dx/dt$$
$$p = m\gamma v$$
$$p^0 = m\gamma$$
$$F = m\gamma d(\gamma v)/dt$$
$$\gamma = (1 - v^2/c^2)^{-\frac{1}{2}}$$

for one spatial dimension with spatial force F. We consider the case of constant F.

It is useful to define a variable z with

$$z = \gamma v \tag{5.4}$$

and to let the velocity of light be a variable quantity v_c that is the speed of light in the vacuum. As a result we find

$$\gamma = (1 - v^2/v_c^2)^{-\frac{1}{2}} \tag{5.5}$$

and the resulting expressions:

$$z^2 = v^2(1 + z^2/v_c^2)$$
$$v^2 = z^2/(1 + z^2/v_c^2)$$
$$\gamma = (1 + z^2/v_c^2)^{1/2} \tag{5.6}$$

Then

$$F = m\gamma d(\gamma v)/dt = m\gamma dz/dt$$
$$= m(1 + z^2/v_c^2)^{1/2}dz/dt$$

We solve for the time dependence with

$$\int F dt = m\int (1 + z^2/v_c^2)^{1/2}dz \tag{5.7}$$

Using

$$\sinh(\theta) = z/v_c \tag{5.8}$$
$$\cosh(\theta) = (1 + z^2/v_c^2)^{1/2} \tag{5.9}$$

we find

$$dz = v_c \cosh\theta \, d\theta \tag{5.10}$$

and

$$\int (1 + z^2/c^2)^{1/2}dz = c \int \cosh^2(\theta) \, d\theta \tag{5.11}$$

As a result

$$F(t - t_0) = mv_c(\theta/2 + \tfrac{1}{4}\sinh(2\theta)) = mv_c[\tfrac{1}{2}\operatorname{arcsinh}(z/v_c) + \tfrac{1}{4}\sinh(2\theta)] \tag{5.12}$$

using

$$\sinh(2\theta) = 2(z/v_c)(1 + z^2/v_c^2)^{1/2} = 2\, v\gamma^2/v_c \tag{5.13}$$

Thus

$$t = (mv_c/2F)[\operatorname{arcsinh}(\gamma v/v_c) + \gamma^2 v/v_c] + t_0 \tag{5.14}$$

The time interval is related to the invariant space-time interval by eq 5.3. Thus the invariant time interval is

$$\tau = mv_c/(2F)\,[\operatorname{arcsinh}(\gamma v/v_c)/\gamma + \gamma v/v_c] + \tau_0 \tag{5.15}$$

We see that larger values of v_c give larger accelerations and velocities per unit time. This fact will motivate our goal of using sling shot gravity boosts in Second Kind White Hole regions. There are also possibilities in ultra-high energy colliding beam accelerators.

5.2 Dynamics for Velocities greater than the Speed of Light

When the speed of light is exceeded, particle (starship) the motion has a slightly different form. Defining

$$i\gamma' = (1 - v^2/v_c^2)^{-1/2} = \gamma \tag{5.16}$$
$$\gamma' = (v^2/v_c^2 - 1)^{-1/2}$$

and

$$z = v\gamma' \tag{5.17}$$

we find

$$z^2(v^2/v_c^2 - 1) = v^2 \tag{5.18}$$

and thus

$$v^2 = z^2/(z^2/v_c^2 - 1) \tag{5.19}$$
$$\gamma = (1 - v^2/v_c^2)^{-\frac{1}{2}} \tag{5.20}$$
$$= -i(z^2/v_c^2 - 1)^{\frac{1}{2}}$$

$$F = m\gamma d(\gamma v)/dt = m\gamma d(i\gamma' v)/dt \tag{5.21}$$
$$= m(z^2/v_c^2 - 1)^{\frac{1}{2}}dz/dt$$

Then

$$\int F dt = m\int (z^2/v_c^2 - 1)^{\frac{1}{2}}dz \tag{5.22}$$

Using

$$\cosh(\theta) = z/v_c \tag{5.23}$$
$$\sinh(\theta) = (z^2/v_c^2 - 1)^{\frac{1}{2}} \tag{5.24}$$

we find

$$dz = v_c \sinh\theta \, d\theta \tag{5.25}$$

and

$$\int (z^2/v_c^2 - 1)^{\frac{1}{2}}dz = v_c \int \sinh^2(\theta) \, d\theta \tag{5.26}$$

As a result

$$F(t - t_0) = mv_c(-\theta/2 + \tfrac{1}{4} \sinh(2\theta)) \tag{5.27}$$
$$= mv_c[-\tfrac{1}{2} \operatorname{arcsinh}(z/v_c) + \tfrac{1}{4} \sinh(2\theta)]$$

using

$$\sinh(2\theta) = 2\sinh(\theta)\cosh(2\theta)$$
$$= 2(z^2/v_c^2 - 1)^{\frac{1}{2}}(z/v_c) = -2 \gamma'^2 v/v_c \tag{5.28}$$

Thus

$$t = -(mv_c/2F)[\operatorname{arcsinh}(\gamma' v/v_c) + \gamma'^2 v/v_c] + t_0 \tag{5.29}$$

The invariant interval is

$$\tau = t/\gamma' = (mv_c/2F) [\gamma' v/v_c + \operatorname{arsinh}(\gamma' v/v_c)/\gamma'] \tag{5.30}$$

6. Superluminal Extension of Newton's Second Law of Mechanics

Comparing the expressions for the invariant interval in chapter 5 we find that they are the same up to a minus sign. The origin of the minus sign for superluminal dynamics is in the form of the dynamic equation, eq. 5.2, for subluminal motion extrapolated directly to superluminal motion in eq. 5.21:

$$F = m\gamma d(\gamma v)/dt = -\, m\gamma' d(\gamma' v)/dt \qquad (6.1)$$

Newton's Laws of Mechanics implements the rule:

The velocity in the direction of the force increases with time.

We see eq. 6.1 clearly violates that rule.
Therefore we suggest superluminal dynamics should implement the rule:[8]

$$F = \, m\gamma' d(\gamma' v)/dt = m|\gamma| d(|\gamma| v)/dt \qquad (6.2)$$

If this rule is accepted then the modified section 5.2 leads to the superluminal solution:

$$t \, = (mv_c/2F)[\mathrm{arcsinh}(\gamma' v/v_c) + \gamma'^2 v/v_c] + t_0 \qquad (5.29')$$

and the invariant interval

$$\tau = t/\gamma' = \, (mv_c/2F)\,[\gamma' v/v_c + \mathrm{arsinh}(\gamma' v/v_c)/\gamma'] \qquad (5.30')$$

As a result the subluminal and superluminal invariant interval expression is:

$$\tau = t/|\gamma| = (mv_c/2F)\,[|\gamma| v/v_c + \mathrm{arsinh}(|\gamma| v/v_c)/|\gamma|] \qquad (6.3)$$

where $|\gamma|$ is the absolute value of γ. Note: the factor $(mv_c/2F)$ sets the time scale.
Chapter 5, and the above discussion, lead to the covariant superluminal and subluminal force law – the proper relativistic extension of Newton's Second Law:

$$F^{\mu} = \, m|\gamma| d(|\gamma| v^{\mu})/dt \qquad (6.4)$$

---

[8] This change, which is equivalent to setting $F \rightarrow -F$, is a change in the effective force and potential for superluminal dynamics seen in chapter 4. It is consistent with the change in the Second Kind White Hole gravitational field from repulsive to attractive for superluminal dynamics.

6.1 Variation of Invariant Time vs. Velocity as a Function of the Effective Speed of Light v_c

A study of Variation of Invariant Time vs. Velocity as a Function of the Effective Speed of Light v_c reveals:

1. The acceleration is reduced if the effective *speed of light* is less than the vacuum speed of light c: . $v_c < c$.

2. The acceleration is increased if the effective *speed of light* is more than the vacuum speed of light c: $v_c > c$.

Consequently the use of Second Kind White Holes improves the acceleration of starships, their engines, and also the acceleration of particles in accelerators.

7. Second Kind White Hole Starship Engines

This engine sends a thrust stream though a small black hole dominated region to propel the input stream with much higher acceleration.

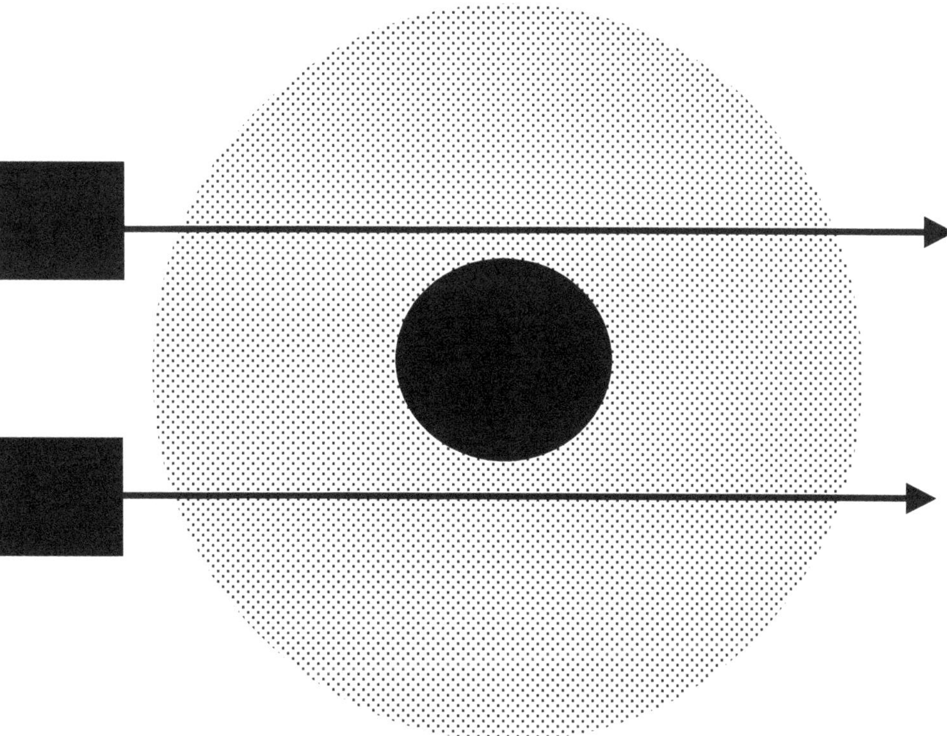

Figure 7.1. Schematic for a Second Kind White Hole starship engine. Input streams generated in the input boxes. The streams more rapidly accelerate through the field of the hole reaching superluminal speed to provide thrust for the starship.

8. Second Kind White Hole Particle Accelerator

This accelerator would cause streams of particles to collide at extreme energies far beyond current colliding beam accelerator energies.

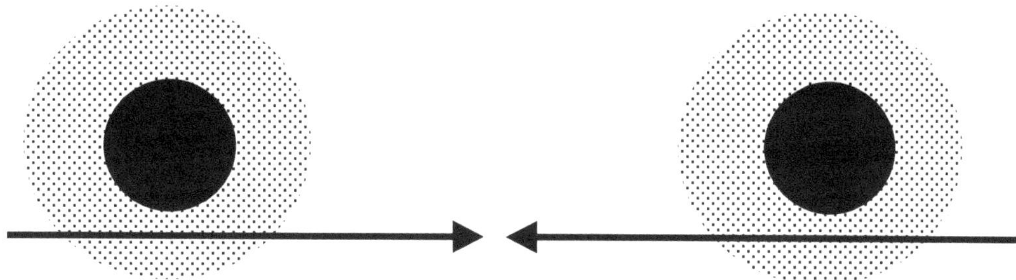

Figure 8.1. Schematic for a Second Kind White Hole colliding beam accelerator. Input streams are accelerated through the hole fields reaching very high energies.

9. The Generation of White Holes

Starship travel may benefit from a gravity assist due to a Second Kind White Hole if one is nearby. However, starship engines and colliding beam accelerators require mini Second Kind White Holes. This chapter presents a plausible scenario for the creation of Second Kind White Holes if certain major technological problems could be overcome. These problems are unlikely to be resolved in the near future.

9.1 Creation of a Hole

The goal is to create a macroscopic mass that is one of a Black Hole, a White Hole or a Second Kind White Hole. A Second Kind White Hole is most desirable in view of the applications described earlier.

We start with the possibility of a nine quark metastable fermion, which we have called[9] *quarkium-9*. We view it as created from the extreme compression of a tritium atom. Tritium is a bound state of 2 neutrons and a proton, which amounts to 5 d quarks and 4 u quarks. Ignoring tritium electrons it has electric charge +1. The quarkium-9 may be created by inertial compression with laser beams much like the beams being used in fusion energy studies. See Fig. 9.1. The energy and compression of these beams is clearly much beyond current fusion energy reactor studies. This effort is thus for the future.

9.2 Macroscopic Quarkium-9 Mass

Current inertial compression laser studies use very low density targets. A massive "Hole" candidate would require a "massive" amount of quarkium-9. Thus the quarkium-9 particles must be relatively long-lived to be assembled into a macroscopic mass.

The problem of the nature of the "Hole" also arises. How to choose the type of the Hole is currently not known. Whether it can be chosen is also an open question. One can hope that the manner in which the Hole is created may influence the type of Hole created. One could, in principle, manipulate the creation process by the choice of laser energies and the assembly time as well as manipulating the electromagnetic fields surrounding the Hole.

9.3 Positioning Quarkium "Holes"

The positioning of a created "Hole" is also an issue. It seems the use of electromagnetic fields could move the Hole into a position and then "lock" it in place.

---

[9] Blaha (2024b). There is a suggestion that such particles may be quasi-stable.

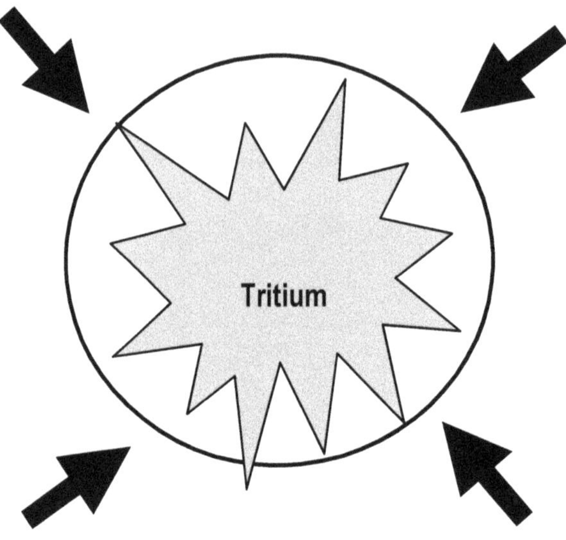

Figure 9.1. Schematic diagram of an imploding tritium atom being compressed into a quarkium-9 particle.

REFERENCES

Akhiezer, N. I., Frink, A. H. (tr), 1962, *The Calculus of Variations* (Blaisdell Publishing, New York, 1962).

Bjorken, J. D., Drell, S. D., 1964, *Relativistic Quantum Mechanics* (McGraw-Hill, New York, 1965).

Bjorken, J. D., Drell, S. D., 1965, *Relativistic Quantum Fields* (McGraw-Hill, New York, 1965).

Blaha, S., 1995, *C++ for Professional Programming* (International Thomson Publishing, Boston, 1995).

_______, 1998, *Cosmos and Consciousness* (Pingree-Hill Publishing, Auburn, NH, 1998 and 2002).

_______, 2002, *A Finite Unified Quantum Field Theory of the Elementary Particle Standard Model and Quantum Gravity Based on New Quantum Dimensions™ & a New Paradigm in the Calculus of Variations* (Pingree-Hill Publishing, Auburn, NH, 2002).

_______, 2004, *Quantum Big Bang Cosmology: Complex Space-time General Relativity, Quantum Coordinates™ Dodecahedral Universe, Inflation, and New Spin 0, ½, 1 & 2 Tachyons & Imagyons* (Pingree-Hill Publishing, Auburn, NH, 2004).

_______, 2005a, *Quantum Theory of the Third Kind: A New Type of Divergence-free Quantum Field Theory Supporting a Unified Standard Model of Elementary Particles and Quantum Gravity based on a New Method in the Calculus of Variations* (Pingree-Hill Publishing, Auburn, NH, 2005).

_______, 2005b, *The Metatheory of Physics Theories, and the Theory of Everything as a Quantum Computer Language* (Pingree-Hill Publishing, Auburn, NH, 2005).

_______, 2005c, *The Equivalence of Elementary Particle Theories and Computer Languages: Quantum Computers, Turing Machines, Standard Model, Superstring Theory, and a Proof that Gödel's Theorem Implies Nature Must Be Quantum* (Pingree-Hill Publishing, Auburn, NH, 2005).

_______, 2006a, *The Foundation of the Forces of Nature* (Pingree-Hill Publishing, Auburn, NH, 2006).

_______, 2006b, *A Derivation of ElectroWeak Theory based on an Extension of Special Relativity; Black Hole Tachyons; & Tachyons of Any Spin.* (Pingree-Hill Publishing, Auburn, NH, 2006).

_______, 2007a, *Physics Beyond the Light Barrier: The Source of Parity Violation, Tachyons, and A Derivation of Standard Model Features* (Pingree-Hill Publishing, Auburn, NH, 2007).

_______, 2007b, *The Origin of the Standard Model: The Genesis of Four Quark and Lepton Species, Parity Violation, the ElectroWeak Sector, Color SU(3), Three Visible Generations of Fermions, and One Generation of Dark Matter with Dark Energy* (Pingree-Hill Publishing, Auburn, NH, 2007).

_______, 2008a, *A Direct Derivation of the Form of the Standard Model From GL(16)* (Pingree-Hill Publishing, Auburn, NH, 2008).

_______, 2008b, *A Complete Derivation of the Form of the Standard Model With a New Method to Generate Particle Masses Second Edition* (Pingree-Hill Publishing, Auburn, NH, 2008)

_______, 2009, *The Algebra of Thought & Reality: The Mathematical Basis for Plato's Theory of Ideas, and Reality Extended to Include A Priori Observers and Space-Time Second Edition* (Pingree-Hill Publishing, Auburn, NH, 2009).

_______, 2009a, *Bright Stars, Bright Universe* (Pingree-Hill Publishing, Auburn, NH, 2008)

_______, 2010a, *Operator Metaphysics: A New Metaphysics Based on a New Operator Logic and a New Quantum Operator Logic that Lead to a Mathematical Basis for Plato's Theory of Ideas and Reality* (Pingree-Hill Publishing, Auburn, NH, 2010).

_______, 2010b, *The Standard Model's Form Derived from Operator Logic, Superluminal Transformations and GL(16)* (Pingree-Hill Publishing, Auburn, NH, 2010).

_______, 2010c, *SuperCivilizations: Civilizations as Superorganisms* (McMann-Fisher Publishing, Auburn, NH, 2010).

_______, 2011a, *21st Century Natural Philosophy Of Ultimate Physical Reality* (McMann-Fisher Publishing, Auburn, NH, 2011).

_______, 2011b, *All the Universe! Faster Than Light Tachyon Quark Starships & Particle Accelerators with the LHC as a Prototype Starship Drive Scientific Edition* (Pingree-Hill Publishing, Auburn, NH, 2011).

_______, 2011c, *From Asynchronous Logic to The Standard Model to Superflight to the Stars* (Blaha Research, Auburn, NH, 2011).

_______, 2012a, *From Asynchronous Logic to The Standard Model to Superflight to the Stars volume 2: Superluminal CP and CPT, U(4) Complex General Relativity and The Standard Model, Complex Vierbein General Relativity, Kinetic Theory, Thermodynamics* (Blaha Research, Auburn, NH, 2012).

______, 2012b, *Standard Model Symmetries, And Four And Sixteen Dimension Complex Relativity; The Origin Of Higgs Mass Terms* (Blaha Reasearch, Auburn, NH, 2012).

______, 2013a, *Multi-Stage Space Guns, Micro-Pulse Nuclear Rockets, and Faster-Than-Light Quark-Gluon Ion Drive Starships* (Blaha Research, Auburn, NH, 2013).

______, 2013b, *The Bridge to Dark Matter; A New Sibling Universe; Dark Energy; Inflatons; Quantum Big Bang; Superluminal Physics; An Extended Standard Model Based on Geometry* (Blaha Reasearch, Auburn, NH, 2013).

______, 2014a, *Universes and Megaverses: From a New Standard Model to a Physical Megaverse; The Big Bang; Our Sibling Universe's Wormhole; Origin of the Cosmological Constant, Spatial Asymmetry of the Universe, and its Web of Galaxies; A Baryonic Field between Universes and Particles; Megaverse Extended Wheeler-DeWitt Equation* (Blaha Reasearch, Auburn, NH, 2014).

______, 2014b, *All the Megaverse! Starships Exploring the Endless Universes of the Cosmos Using the Baryonic Force* (Blaha Research, Auburn, NH, 2014).

______, 2014c, *All the Megaverse! II Between Megaverse Universes: Quantum Entanglement Explained by the Megaverse Coherent Baryonic Radiation Devices – PHASERs Neutron Star Megaverse Slingshot Dynamics Spiritual and UFO Events, and the Megaverse Microscopic Entry into the Megaverse* (Blaha Research, Auburn, NH, 2014).

______, 2015a, *PHYSICS IS LOGIC PAINTED ON THE VOID: Origin of Bare Masses and The Standard Model in Logic, U(4) Origin of the Generations, Normal and Dark Baryonic Forces, Dark Matter, Dark Energy, The Big Bang, Complex General Relativity, A Megaverse of Universe Particles* (Blaha Research, Auburn, NH, 2015).

______, 2015b, *PHYSICS IS LOGIC Part II: The Theory of Everything, The Megaverse Theory of Everything, U(4)$\otimes$U(4) Grand Unified Theory (GUT), Inertial Mass = Gravitational Mass, Unified Extended Standard Model and a New Complex General Relativity with Higgs Particles, Generation Group Higgs Particles* (Blaha Research, Auburn, NH, 2015).

______, 2015c, *The Origin of Higgs ("God") Particles and the Higgs Mechanism: Physics is Logic III, Beyond Higgs – A Revamped Theory With a Local Arrow of Time, The Theory of Everything Enhanced, Why Inertial Frames are Special, Universes of the Mind* (Blaha Research, Auburn, NH, 2015).

______, 2015d, *The Origin of the Eight Coupling Constants of The Theory of Everything: U(8) Grand Unified Theory of Everything (GUTE), S^8 Coupling Constant Symmetry, Space-Time Dependent Coupling Constants, Big Bang Vacuum Coupling Constants, Physics is Logic IV* (Blaha Research, Auburn, NH, 2015).

______, 2016a, *New Types of Dark Matter, Big Bang Equipartition, and A New U(4) Symmetry in the Theory of Everything: Equipartition Principle for Fermions, Matter is 83.33% Dark, Penetrating the Veil of the Big Bang, Explicit QFT Quark Confinement and Charmonium, Physics is Logic V* (Blaha Research, Auburn, NH, 2016).

______, 2016b, *The Periodic Table of the 192 Quarks and Leptons in The Theory of Everything: The U(4) Layer Group, Physics is Logic VI* (Blaha Research, Auburn, NH, 2016).

______, 2016c, *New Boson Quantum Field Theory, Dark Matter Dynamics, Dark Matter Fermion Layer Mixing, Genesis of Higgs Particles, New Layer Higgs Masses, Higgs Coupling Constants, Non-Abelian Higgs Gauge Fields, Physics is Logic VII* (Blaha Research, Auburn, NH, 2016).

______, 2016d, *Unification of the Strong Interactions and Gravitation: Quark Confinement Linked to Modified Short-Distance Gravity; Physics is Logic VIII* (Blaha Research, Auburn, NH, 2016).

______, 2016e, *MoND: Unification of the Strong Interactions and Gravitation II, Quark Confinement Linked to Large-Scale Gravity, Physics is Logic IX* (Blaha Research, Auburn, NH, 2016).

______, 2016f, *CQ Mechanics: A Unification of Quantum & Classical Mechanics, Quantum/Semi-Classical Entanglement, Quantum/Classical Path Integrals, Quantum/Classical Chaos* (Blaha Research, Auburn, NH, 2016).

______, 2016g, *GEMS Unified Gravity, ElectroMagnetic and Strong Interactions: Manifest Quark Confinement, A Solution for the Proton Spin Puzzle, Modified Gravity on the Galactic Scale* (Pingree Hill Publishing, Auburn, NH, 2016).

______, 2016h, *Unification of the Seven Boson Interactions based on the Riemann-Christoffel Curvature Tensor* (Pingree Hill Publishing, Auburn, NH, 2016).

______, 2017a, *Unification of the Eleven Boson Interactions based on 'Rotations of Interactions'* (Pingree Hill Publishing, Auburn, NH, 2017).

______, 2017b, *The Origin of Fermions and Bosons, and Their Unification* (Pingree Hill Publishing, Auburn, NH, 2017).

______, 2017c, *Megaverse: The Universe of Universes* (Pingree Hill Publishing, Auburn, NH, 2017).

______, 2017d, *SuperSymmetry and the Unified SuperStandard Model* (Pingree Hill Publishing, Auburn, NH, 2017).

______, 2017e, *From Qubits to the Unified SuperStandard Model with Embedded SuperStrings: A Derivation* (Pingree Hill Publishing, Auburn, NH, 2017).

______, 2017f, *The Unified SuperStandard Model in Our Universe and the Megaverse: Quarks, … ,* (Pingree Hill Publishing, Auburn, NH, 2017).

______, 2018a, *The Unified SuperStandard Model and the Megaverse SECOND EDITION A Deeper Theory based on a New Particle Functional Space that Explicates Quantum Entanglement Spookiness (Volume 1)* (Pingree Hill Publishing, Auburn, NH, 2018).

______, 2018b, *Cosmos Creation: The Unified SuperStandard Model, Volume 2, SECOND EDITION* (Pingree Hill Publishing, Auburn, NH, 2018).

______, 2018c, *God Theory (*Pingree Hill Publishing, Auburn, NH, 2018).

______, 2018d, *Immortal Eye: God Theory: Second Edition* (Pingree Hill Publishing, Auburn, NH, 2018).

______, 2018e, *Unification of God Theory and Unified SuperStandard Model THIRD EDITION* (Pingree Hill Publishing, Auburn, NH, 2018).

______, 2019a, *Calculation of: QED α = 1/137, and Other Coupling Constants of the Unified SuperStandard Theory* (Pingree Hill Publishing, Auburn, NH, 2019).

______, 2019b, *Coupling Constants of the Unified SuperStandard Theory SECOND EDITION* (Pingree Hill Publishing, Auburn, NH, 2019).

______, 2019c, *New Hybrid Quantum Big_Bang–Megaverse_Driven Universe with a Finite Big Bang and an Increasing Hubble Constant* (Pingree Hill Publishing, Auburn, NH, 2019).

______, 2019d, *The Universe, The Electron and The Vacuum* (Pingree Hill Publishing, Auburn, NH, 2019).

______, 2019e, *Quantum Big Bang – Quantum Vacuum Universes (Particles)* (Pingree Hill Publishing, Auburn, NH, 2019).

______, 2019f, *The Exact QED Calculation of the Fine Structure Constant Implies ALL 4D Universes have the Same Physics/Life Prospects* (Pingree Hill Publishing, Auburn, NH, 2019).

______, 2019g, *Unified SuperStandard Theory and the SuperUniverse Model: The Foundation of Science* (Pingree Hill Publishing, Auburn, NH, 2019).

______, 2020a, *Quaternion Unified SuperStandard Theory (The QUeST) and Megaverse Octonion SuperStandard Theory (MOST)* (Pingree Hill Publishing, Auburn, NH, 2020).

______, 2020b, *United Universes Quaternion Universe - Octonion Megaverse* (Pingree Hill Publishing, Auburn, NH, 2020).

______, 2020c, *Unified SuperStandard Theories for Quaternion Universes & The Octonion Megaverse* (Pingree Hill Publishing, Auburn, NH, 2020).

______, 2020d, *The Essence of Eternity: Quaternion & Octonion SuperStandard Theories* (Pingree Hill Publishing, Auburn, NH, 2020).

______, 2020e, *The Essence of Eternity II* (Pingree Hill Publishing, Auburn, NH, 2020).

______, 2020f, *A Very Conscious Universe* (Pingree Hill Publishing, Auburn, NH, 2020).

______, 2020g, *Hypercomplex Universe* (Pingree Hill Publishing, Auburn, NH, 2020).

______, 2020h, *Beneath the Quaternion Universe* (Pingree Hill Publishing, Auburn, NH, 2020).

______, 2020i, *Why is the Universe Real? From Quaternion & Octonion to Real Coordinates* (Pingree Hill Publishing, Auburn, NH, 2020).

______, 2020j, *The Origin of Universes: of Quaternion Unified SuperStandard Theory (QUeST); and of the Octonion Megaverse (UTMOST)* (Pingree Hill Publishing, Auburn, NH, 2020).

______, 2020k, *The Seven Spaces of Creation: Octonion Cosmology* (Pingree Hill Publishing, Auburn, NH, 2020).

______, 2020l, *From Octonion Cosmology to the Unified SuperStandard Theory of Particles* (Pingree Hill Publishing, Auburn, NH, 2020).

______, 2021a, *Pioneering the Cosmos* (Pingree Hill Publishing, Auburn, NH, 2021).

______, 2021b, *Pioneering the Cosmos II* (Pingree Hill Publishing, Auburn, NH, 2021).

______, 2021c, *Beyond Octonion Cosmology* (Pingree Hill Publishing, Auburn, NH, 2021).

______, 2021d, *Universes are Particles* (Pingree Hill Publishing, Auburn, NH, 2021).

______, 2021e, *Octonion-like dna-based life, Universe expansion is decay, Emerging New Physics* (Pingree Hill Publishing, Auburn, NH, 2021).

______, 2021f, *The Science of Creation New Quantum Field Theory of Spaces* (Pingree Hill Publishing, Auburn, NH, 2021).

______, 2021g, *Quantum Space Theory With Application to Octonion Cosmology & Possibly To Fermionic Condensed Matter* (Pingree Hill Publishing, Auburn, NH, 2021).

______, 2021h, *21st Century Natural Philosophy of Octonion Cosmology , and Predestination, Fate, and Free Will* (Pingree Hill Publishing, Auburn, NH, 2021).

______, 2021i, *Beyond Octonion Cosmology II : Origin of the Quantum; A New Generalized Field Theory (GiFT); A Proof of the Spectrum of Universes; Atoms in Higher Universes* (Pingree Hill Publishing, Auburn, NH, 2021).

______, 2021j, *Integration of General Relativity and Quantum Theory: Octonion Cosmology, GiFT, Creation/Annihilation Spaces CASe, Reduction of Spaces to a Few Fermions and Symmetries in Fundamental Frames* (Pingree Hill Publishing, Auburn, NH, 2021).

______, 2022a, *New View of Octonion Cosmology Based on the Unification of General Relativity and Quantum Theory* (Pingree Hill Publishing, Auburn, NH, 2022).

______, 2022b, *The Dust Beneath Hypercomplex Cosmology* (Pingree Hill Publishing, Auburn, NH, 2022).

______, 2022c, *Passing Through Nature to Eternity: ProtoCosmos, HyperCosmos, Unified SuperStandard Theory* (Pingree Hill Publishing, Auburn, NH, 2022).

______, 2022d, *HyperCosmos Fractionation and Fundamental Reference Frame Based Unification: Particle Inner Space Basis of Parton and Dual Resonance Models* (Pingree Hill Publishing, Auburn, NH, 2022).

______, 2022e, *A New UniDimension ProtoCosmos and SuperString F-Theory Relation to the HyperCosmos* (Pingree Hill Publishing, Auburn, NH, 2022).

______, 2022f, *The Cosmic Panorama: ProtoCosmos, HyperCosmos,Unified SuperStandard Theory (UST) Derivation* (Pingree Hill Publishing, Auburn, NH, 2022).

______, 2022g, *Ultimate Origin: ProtoCosmos and HyperCosmos* (Pingree Hill Publishing, Auburn, NH, 2022).

______, 2023a, *UltraUnification and the Generation of the Cosmos* (Pingree Hill Publishing, Auburn, NH, 2023).

______, 2023b, *God and and Cosmos Theory* (Pingree Hill Publishing, Auburn, NH, 2023).

______, 2023c, *A New Completely Geometric SU(8) Cosmos Theory; New PseudoFermion Fields; Fibonacci-like Dimension Arrays; Ramsey Number Approximation* (Pingree Hill Publishing, Auburn, NH, 2023).

______, 2023d, *Newton's Apple is Now the Fermion* (Pingree Hill Publishing, Auburn, NH, 2023).

______, 2023e,*Cosmos Theory: The Sub-Particle Gambol Model* (Pingree Hill Publishing, Auburn, NH, 2023).

______, 2024a, *Cosmos-Universe-Particle-Gambol Theory* (Pingree Hill Publishing, Auburn, NH, 2024).

______, 2024b, *Fractal Cosmos Theory* (Pingree Hill Publishing, Auburn, NH, 2024).

______, 2024c, *Fractal Cosmic Curve: Tensor-Based CosmosTheory* (Pingree Hill Publishing, Auburn, NH, 2024).

______, 2024d, *The Eternal Form of Cosmos Theory (*Pingree Hill Publishing, Auburn, NH, 2024).

______, 2024e, *The Eternal Form of Cosmos Theory Third Edition (*Pingree Hill Publishing, Auburn, NH, 2024).

______, 2024f, *Fundamental Constants of Cosmos Theory and The Standard Model* (Pingree Hill Publishing, Auburn, NH, 2024).

______, 2024g, *Quark, Lepton, W and Z Masses of Cosmos Theory and The Standard Model* (Pingree Hill Publishing, Auburn, NH, 2024).

______, 2024h, *Geometric Cosmos Geometric Universe* (Pingree Hill Publishing, Auburn, NH, 2024).

______, 2024i, *Particles and Universes of Cosmos Theory* (Pingree Hill Publishing, Auburn, NH, 2024).

______, 2024j, *Unification of the Subluminal and the Superluminal in Cosmos Theory* (Pingree Hill Publishing, Auburn, NH, 2024).

______, 2024k, *The Dawn of Dynamic Cosmos Dimension Arrays* (Pingree Hill Publishing, Auburn, NH, 2024).

______, 2024l, *Structure and Dynamics of Cosmos Theory and the Unified SuperStandard Theory* (Pingree Hill Publishing, Auburn, NH, 2024).

Eddington, A. S., 1952, *The Mathematical Theory of Relativity* (Cambridge University Press, Cambridge, U.K., 1952).

Fant, Karl M., 2005, *Logically Determined Design: Clockless System Design With NULL Convention Logic* (John Wiley and Sons, Hoboken, NJ, 2005).

Feinberg, G. and Shapiro, R., 1980, *Life Beyond Earth: The Intelligent Earthlings Guide to Life in the Universe* (William Morrow and Company, New York, 1980).

Gelfand, I. M., Fomin, S. V., Silverman, R. A. (tr), 2000, *Calculus of Variations* (Dover Publications, Mineola, NY, 2000).

Giaquinta, M., Modica, G., Souchek, J., 1998, *Cartesian Coordinates in the Calculus of Variations* Volumes I and II (Springer-Verlag, New York, 1998).

Giaquinta, M., Hildebrandt, S., 1996, *Calculus of Variations* Volumes I and II (Springer-Verlag, New York, 1996).

Gradshteyn, I. S. and Ryzhik, I. M., 1965, *Table of Integrals, Series, and Products* (Academic Press, New York, 1965).

Heitler, W., 1954, *The Quantum Theory of Radiation* (Claendon Press, Oxford, UK, 1954).

Huang, Kerson, 1992, *Quarks, Leptons & Gauge Fields* 2nd *Edition* (World Scientific Publishing Company, Singapore, 1992).

Jost, J., Li-Jost, X., 1998, *Calculus of Variations* (Cambridge University Press, New York, 1998).

Kaku, Michio, 1993, *Quantum Field Theory*, (Oxford University Press, New York, 1993).

Kirk, G. S. and Raven, J. E., 1962, *The Presocratic Philosophers* (Cambridge University Press, New York, 1962).

Landau, L. D. and Lifshitz, E. M., 1987, *Fluid Mechanics 2nd Edition*, (Pergamon Press, Elmsford, NY, 1987).

Rescher, N., 1967, *The Philosophy of Leibniz* (Prentice-Hall, Englewood Cliffs, NJ, 1967).

Riesz, Frigyes and Sz.-Nagy, Béla, 1990, *Functional Analysis* (Dover Publications, New York, 1990).

Sakurai, J. J., 1964, *Invariance Principles and Elementary Particles* (Princeton University Press, Princeton, NJ, 1964).

Weinberg, S., 1972, *Gravitation and Cosmology* (John Wiley and Sons, New York, 1972).

Weinberg, S., 1995, *The Quantum Theory of Fields Volume I* (Cambridge University Press, New York, 1995).

About the Author

Stephen Blaha is a well-known Physicist and Man of Letters with interests in Science, Society and civilization, the Arts, and Technology. He had an Alfred P. Sloan Foundation scholarship in college. He received his Ph.D. in Physics from Rockefeller University. He has served on the faculties of several major universities. He was also a Member of the Technical Staff at Bell Laboratories, a manager at the Boston Globe Newspaper, a Director at Wang Laboratories, and President of Blaha Software Inc. and of Janus Associates Inc. (NH).

Among other achievements he was a co-discoverer of the "r potential" for heavy quark binding developing the first (and still the only demonstrable) non-Aeolian gauge theory with an "r" potential; first suggested the existence of topological structures in superfluid He-3; first proposed Yang-Mills theories would appear in condensed matter phenomena with non-scalar order parameters; first developed a grammar-based formalism for quantum computers and applied it to elementary particle theories; first developed a new form of quantum field theory without divergences (thus solving a major 60 year old problem that enabled a unified theory of the Standard Model and Quantum Gravity without divergences to be developed); first developed a formulation of complex General Relativity based on analytic continuation from real space-time; first developed a generalized non-homogeneous Robertson-Walker metric that enabled a quantum theory of the Big Bang to be developed without singularities at t = 0; first generalized Cauchy's theorem and Gauss' theorem to complex, curved multi-dimensional spaces; received Honorable Mention in the Gravity Research Foundation Essay Competition in 1978; first developed a physically acceptable theory of faster-than-light particles; first derived a composition of extremums method in the Calculus of Variations; first quantitatively suggested that inflationary periods in the history of the universe were not needed; first proved Gödel's Theorem implies Nature must be quantum; provided a new alternative to the Higgs Mechanism, and Higgs particles, to generate masses; first showed how to resolve logical paradoxes including Gödel's Undecidability Theorem by developing Operator Logic and Quantum Operator Logic; first developed a quantitative harmonic oscillator-like model of the life cycle, and interactions, of civilizations; first showed how equations describing superorganisms also apply to civilizations. A recent book shows his theory applies successfully to the past 14 years of history and to *new* archaeological data on Andean and Mayan civilizations as well as Early Anatolian and Egyptian civilizations.

He first developed an axiomatic derivation of the form of The Standard Model from geometry – space-time properties – The Unified SuperStandard Model. It unifies all the known forces of Nature. It also has a Dark Matter sector that includes a Dark ElectroWeak sector with Dark doublets and Dark gauge interactions. It uses quantum coordinates to remove infinities that crop up in most interacting quantum field theories and additionally to remove the infinities that appear in the Big Bang and generate inflationary growth of the universe. It shows gravity has a MOND-like form without sacrificing Newton's Laws. It relates the interactions of the MOND-like sector of gravity with the r-potential of Quark Confinement. The axioms of the theory lead to the question of their origin. We suggest in the preceding edition of this book it can be attributed to an entity with God-like properties. We explore these properties in "God Theory" and show they predict that the Cosmos exists forever although individual universes (or incarnations of our universe) "come and go." Several other important results emerge from God Theory such a functionally triune God. The Unified SuperStandard Theory has many other important parts described in the Current Edition of *The Unified SuperStandard Theory* and expanded in subsequent volumes.

Blaha has had a major impact on a succession of elementary particle theories: his Ph.D. thesis (1970), and papers, showed that quantum field theory calculations to all orders in ladder approximations could not give scaling deep inelastic electron-nucleon scattering. He later showed the eigenvalue equation for the fine structure constant α in Johnson-Baker-Willey QED had a zero at $\alpha = 1$ not 1/137 by solving the Schwinger-Dyson equations to all orders in an approximation that agreed with exact results to 4th order in α thus ending interest in this theory. In 1979 at Prof. Ken Johnson's (MIT) suggestion he calculated the proton-neutron mass difference in the MIT bag model and found the result had the wrong sign reducing interest in the bag model. These results all appear in Physical Review papers. In the 2000's he repeatedly pointed out the shortcomings of SuperString theory and showed that The Standard Model's form could be derived from space-time geometry by an extension of Lorentz transformations to faster than light transformations. This deeper space-time basis greatly increases the possibility that it is part of THE fundamental theory. Recently, Blaha showed that the Weak interactions differed significantly from the Strong, electromagnetic and gravitation interactions in important respects while these interactions had similar features, and suggested that ElectroWeak theory, which is essentially a glued union of the Weak interactions and Electromagnetism, possibly modulo unknown Higgs particle features, be replaced by a unified theory of the other interactions combined with a stand-alone Weak interaction theory. Blaha also showed that, if Charmonium calculations are taken seriously, the Strong interaction coupling constant is only a factor of five larger than the electromagnetic coupling constant, and thus Strong interaction perturbation theory would make sense and yield physically meaningful results.

In graduate school (1965-71) he wrote substantial papers in elementary particles and group theory: The Inelastic E- P Structure Functions in a Gluon Model. Phys. Lett. B40:501-502,1972; Deep-Inelastic E-P Structure Functions In A Ladder Model With Spin 1/2 Nucleons, Phys.Rev. D3:510-523,1971; Continuum Contributions To The Pion Radius, Phys. Rev. 178:2167-2169,1969; Character Analysis of U(N) and SU(N), J. Math. Phys. 10, 2156 (1969); and The Calculation of the Irreducible Characters of the Symmetric Group in Terms of the Compound Characters, (Published as Blaha's Lemma in D. E. Knuth's book: *The Art of Computer Programming Vols. 1 – 4*).

In the early 1980's Blaha was also a pioneer in the development of UNIX for financial, scientific and Internet applications: benchmarked UNIX versions showing that block size was critical for UNIX performance, developing financial modeling software, starting database benchmarking comparison studies, developing Internet-like UNIX networking (1982) and developing a hybrid shell programming technique (1982) that was a precursor to the PERL programming language. He was also the manager of the AT&T ten-year future products development database. His work helped lead to commercial UNIX on computers such as Sun Micros, IBM AIX minis, and Apple computers.

In the 1980's he pioneered the development of PC Desktop Publishing on laser printers and was nominated for three "Awards for Technical Excellence" in 1987 by PC Magazine for PC software products that he designed and developed.

Recently he has developed a theory of Megaverses – actual universes of which our universe is one – with quantum particle-like properties based on the Wheeler-DeWitt equation of Quantum Gravity. He has developed a theory of a baryonic force, which had been conjectured many years ago, and estimated the strength of the force based on discrepancies in measurements of the gravitational constant G. This force, operative in D-dimensional space, can be used to escape from our universe in "uniships" which are the equivalent of the faster-than-light starships proposed in the author's earlier books. Thus travel to other universes, as well as to other stars is possible.

Blaha also considered the complexified Wheeler-DeWitt equation and showed that its limitation to real-valued coordinates and metrics generated a Cosmological Constant in the Einstein equations.

The author has also recently written a series of books on the serious problems of the United States and their solution as well as a book on the decline of Mankind that will follow from current social and genetic trends in Mankind.

In the past twenty years Dr. Blaha has written over 80 books on a wide range of topics. Some recent major works are: *From Asynchronous Logic to The Standard Model to Superflight to the Stars, All the Universe!, SuperCivilizations: Civilizations as Superorganisms, America's Future: an Islamic Surge, ISIS, al Qaeda, World Epidemics, Ukraine, Russia-China Pact, US Leadership Crisis, The Rises and Falls of Man – Destiny – 3000 AD: New Support for a Superorganism MACRO-THEORY of CIVILIZATIONS From CURRENT WORLD TRENDS and NEW Peruvian, Pre-Mayan, Mayan, Anatolian, and Early Egyptian Data, with a Projection to 3000 AD,* and *Mankind in Decline: Genetic Disasters, Human-Animal Hybrids, Overpopulation, Pollution, Global Warming, Food and Water Shortages, Desertification, Poverty, Rising Violence, Genocide, Epidemics, Wars, Leadership Failure.*

He has taught approximately 4,000 students in undergraduate, graduate, and postgraduate corporate education courses primarily in major universities, and large companies and government agencies.

He developed a quantum theory, The Unified SuperStandard Theory (UST), which describes elementary particles in detail without the difficulties of conventional quantum field theory. He found that the internal symmetries of this theory could be exactly derived from an octonion theory called QUeST. He further found that another octonion theory (UTMOST) describes the Megaverse. It can hold QUeST universes such as our own universe. It has an internal symmetry structure which is a superset of the QUeST internal symmetries.

Recently he developed Octonion Cosmology. He replaced it with HyperCosmos theory, which has significantly better features. He developed a fractionalization process for dimensions, particles and symmetry groups. He also described transformation that reduced particles and dimensions to a far more compact form. He also developed a precursor theory ProtoCosmos that leads to the HyperCosmos.

The author showed that space-time and Internal Symmetries can be unified in any of the ten HyperCosmos spaces in their associated HyperUnification spaces. The combined set of HyperUnification spaces enable all HyperCosmos dimensions to be obtained by a General Relativistic transformation from one primordial dimension in the 42 space-time dimension unified HyperUnification space.

At present the author devel;oped the Cosmos Theory that incorporates ProtoCosmos Theory, HyperCosmos Theory, Limos Theory, Second Kind HyperCosmos Theory and HyperUnification Spaces. He has introduced PseudoFermion wave functions and theory, He has related Cosmos Theory to Regge trajectories of spaces, parton theory, Veneziano amplitudes, Fibonacci numbers and Ramsey numbers. He has calculated an approximation to the difficult R(n,n) Ramsey numbers.

He has developed a Gambol Model that successfully accounts for e-p deep inelastic scattering, fundamental particle resonances, hadron scattering, and the inner structure of particles based on confinement through Casimir forces of ideal gambol gases. The Gambol Planckian Distribution was derived.

He has applied the Gambol Model to particles, universes, and the Cosmos of universes. He showed that the Cosmos may have a distribution of 23 universes corresponding to various Cosmos spaces.

Recently he showed that Cosmos Theory follows from the number of independent asymmetric tensors in a dimension r. He also showed the close parallel between the form of γ-matrices and Cosmos Theory dimension arrays. The closeness suggested that dimension arrays have the same importance as γ-matrices for fermions.

He demonstrated that the pressure of fermions within a space of dimension r balances the Casimir vacuum energy force for 18 dimensions. He showed that $2e\pi = 17.02$ marks the critical point where pressure balances Casimir force, which implies r = 18 is the highest dimension Physical Cosmos space. The dimension $2e\pi$ appears to set the approximate dimension for Cosmos spaces with dimension array size $2^{r+4} \cong (17.02/8)^{r+4}. \cong (e\pi/4)^{r+4}. \cong 2.13^{r+4}$.

Recent books have shown that the main features of The Standard Model and the Unified SuperStandard Theory are present in Cosmos Theory.